YOUR KNOWLEDGE HAS VALUE

- We will publish your bachelor's and master's thesis, essays and papers

- Your own eBook and book - sold worldwide in all relevant shops

- Earn money with each sale

Upload your text at www.GRIN.com and publish for free

Bibliographic information published by the German National Library:

The German National Library lists this publication in the National Bibliography; detailed bibliographic data are available on the Internet at http://dnb.dnb.de .

Imprint:

Copyright © 2011 GRIN Verlag, Open Publishing GmbH
Print and binding: Books on Demand GmbH, Norderstedt Germany
ISBN: 978-3-668-11081-6

This book at GRIN:

http://www.grin.com/en/e-book/179642/project-report-on-asuro-robot

James Sutanto, Li Ze Cheng, Shi Juo Jun, Wen Bo, Chen Zhi Ze

Project report on Asuro robot

GRIN Publishing

Xi'an Jiaotong-Liverpool University

THE REPORT OF ASURO

James Sutanto
(Contribution in elaborate the Abstract, Procedure, Troubleshooting, Chart,
Conclusion)

Li Ze Cheng
(Contribution in elaborate the introduction)

Shi Juo Jun
(Contribution in Discussion parts)

Wen Bo
(Contribution in theory and circuit design)

Chen Zhi Ze
(Contribute in results)

Electrical Engineering and Automation

Submit date: 2010.11.22

Asuro Robot

Abstract: *This article illustrates the project about the Asuro Robot, which conducted in Xi'an Jiao Tong Liverpool University. In this study, students required to learn the practical skills in building and troubleshooting a circuit. However, there were several difficulties occurred in the experiment, such as the polarities of the components. Due to these difficulties, students need to be well prepared about the basic of electronic circuit.*

Introduction

Since the beginnings of civilization man has had a fascination for a human-like creation that would assist him. Robots, such as human creatures built in by the electronic devices and created into one. In this experiment, Asuro Robot were chosen because this robot equipped with a diversity of components such as an Atmel AVR RISC-processor, two independently controlled motors, an optical line tracer, six collision detector switches, two odometer-sensors, three indicators LED, an IR-Interface for programming and remote controlled by a PC. These components all have their own specially function. For instance, the line tracer is used to help the robot go through a traced way, the orders given to the robot to make possible ASURO follows the road can be designed and developed by C programming. In order to assemble and install properly the elements of the robot, basic knowledge about soldering is requested and is expected the builder improve his skills.

In this experiment, each participant achieves the project through the team work. On the one hand people can learn how to weld the devices, and how to use computer to write program into the robot. One other hand, people also can learn the skills which are how to communicate with other people and how to coordinate the teammates through the team work.

Theory:

The basic theory of the experiment was Ohm's law: The current flowing in the resistor is proportional to the voltage across the resistor. And the formula is: V=i*R, where V=voltage (V), i=current (A), and R=resistance (Ω).

Also the capacitor was very important: the voltage difference between the two conductors is proportional to the charge and q=C v, therefore i=dq/dt=C dv/dt. The proportionality constant C is called capacitance. The units of Farads (F) =Coulomb/Volt and for two parallel plates: C= εA/d.

$$i(t)=CC\frac{dv(t)}{dt} \qquad v(t)=\frac{1}{C}\int\int_{-\infty}^{t}i(x)dx$$

Experimental method:

The materials use in this project consists of many different type of electronic components. such as resistor, capacitor, LEDs light, and transistor. Furthermore, for the equipment use in this project was solder, multimeter. See appendix for partlists

Circuit Design:

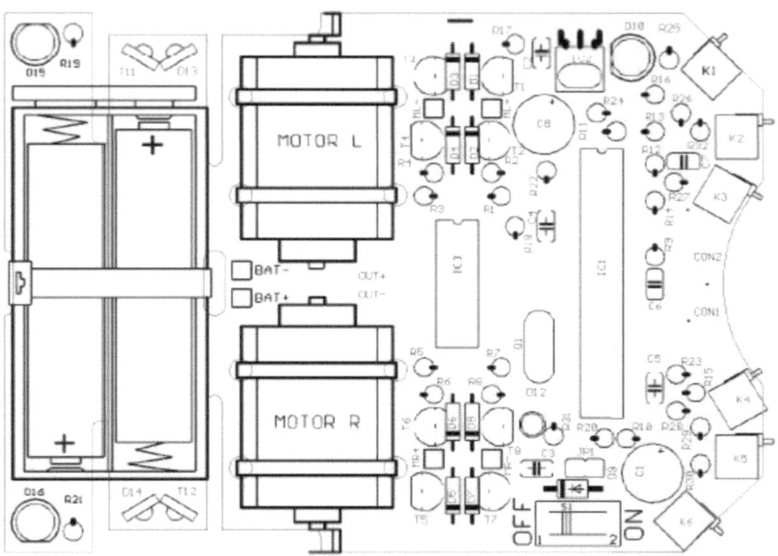

Fig. 1 Component layout

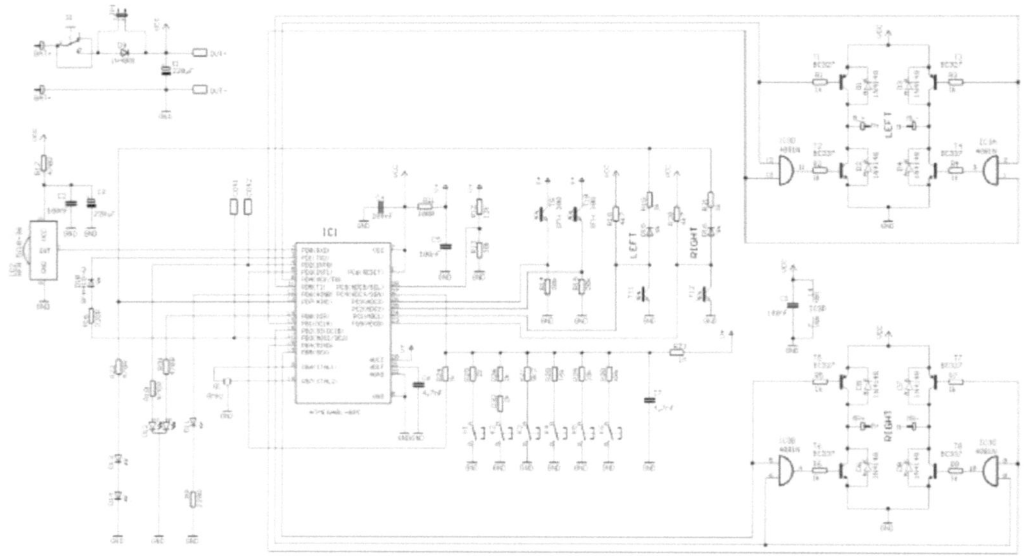

Fig.2 Schematic Diagram of the PCB circuit

Procedure:

The following inserting sequence may be used as a guideline:
- IC1: at first insert only the socket (either one Dual in line 28 pole-socket or two Dual in line
- 14 pole sockets). Pay attention to polarity! A slight asymmetry is a mark for polarity in the socket and in the symbols on the PCB!
- IC3: Again insert only the socket (Dual in line 14 pole-socket). Pay attention to polarity! A slight asymmetry is a mark for polarity in the socket and in the symbols on the PCB!
- K1, K2, K3, K4, K5, K6: Sensor-switches, which must be mounted fl at to the PCB surface!
- Q1; Resonator 8MHz
- D1, D2, D3, D4, D5, D6, D7, D8: Diode 1N4148; Pay attention to polarity!
- D9: 1N4001; Pay attention to polarity!
- JP1: Jumper; Short pins will be soldered; do not yet apply the jumper connector element!
- D12: Dual colored LED, 3 mm diameter, three legs; Pay attention to polarity! (Polarity may differ from part to part, however: the shortest leg must be inserted into the square soldering pad)!
- C2, C3, C4, C5: 100nF Ceramic; Imprinted: 104
- C6, C7: 4,7nF Ceramic; Imprinted: 472
- T1, T3, T5, T7: BC327-40 or BC328-40
- T2, T4, T6, T8: BC337-40 or BC338-40
- R1, R2, R3, R4, R5, R6, R7, R8, R19, R21: 1KΩ 5% (brown, black, red, gold)
- R9, R16: 220Ω 5% (red, red, brown, gold)
- R10, R17, R22, R31: 5% 470Ω (yellow, violet, brown, gold)
- R11: 100Ω 5% (brown, black, brown, gold)
- R12: 12KΩ 1% (brown, red, orange, gold)
- R13: 10KΩ 1% (brown, black, orange, gold)
- R14, R15: 20KΩ 5% (red, black, orange, gold)
- R18, R20: 4,7KΩ 5% (yellow, violet, red, gold)
- R23: 1MΩ 5% (brown, black, green, gold)
- R24: 1KΩ 1% (brown, black, black, brown, brown)
- R25, R26, *R32*: 2KΩ 1% (red, black, black, brown, brown)
- R27: 8,2KΩ 1% (grey, red, black, brown, brown)
- R28: 16KΩ 1% (brown, blue, black, red, brown)
- R29: 33KΩ 1% (orange, orange, black, red, brown)
- R30: 68KΩ 1% (blue, grey, black, red, brown)
- C1, C8: Elco 220µF 10V or higher values; Pay attention to polarity!
- IC2: SFH5110-36 Infrared-receiver-IC, bend the legs with long-nose pliers!
- Pay attention to polarity (the side with dome-shaped curvature must be positioned to the outside)! Caution: electrostatic discharge (ESD) and excessive soldering or heating may damage the part!

- o D10: SFH 415-U IR-LED 5mm; black case; Pay attention to polarity! Case must be settled
  close to PCB.
- o T11, T12: LPT80A, Phototransistor, colorless case;
- o Case must be settled close to PCB; Pay attention to polarity!
- o D13, D14: IRL80A, IR-LED, rosy case;
- o Case must be settled close to PCB; Pay attention to polarity!
- o D15, D16: LED 5 mm red, rosy respectively red case. Pay attention to polarity
- o (Short leg must be inserted at the mark)!
- o S1: On/Off-Switch
- o Three more parts will be needed (which will be used to follow a line), but they will be placed at the bottom side of the PCB and have to soldered from the upper side
- o T9, T10: SFH300, Phototransistor 5 mm; Pay attention to polarity! These components have to be placed at some distance from the PCB.
- o D11: LED 5 mm red, red or reddish case; Pay attention to polarity! (short leg must be
  inserted at the mark)!

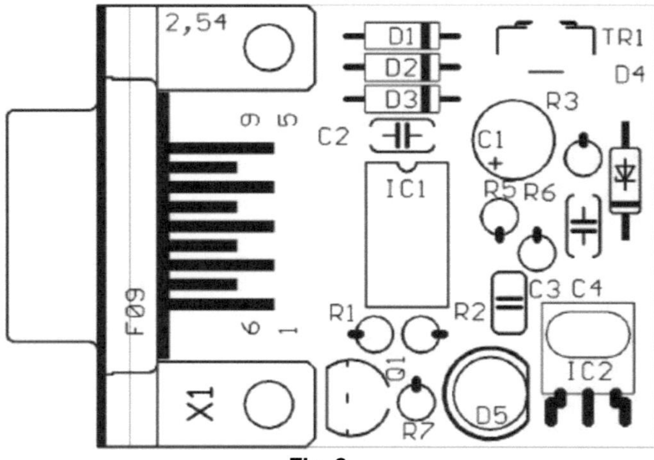

Fig. 3

Assembling the RS232 infrared-transceiver
- o IC1: Initially insert the 8-pole socket. The polarity mark of the (slightly asymmetrical) socket must correspond to the mark in the accompanying symbol on the PCB.
- o D1, D2, D3: 1N4148, pay attention to polarity! Read the imprints of the parts and take care not to interchange with ZPD5.1 or BZX55-C5V1!
- o D4: ZPD5.1 or BZX55-C5V1, pay attention to polarity! Read the imprints of the parts and
- o take care not to interchange with 1N4148!

- D5: SFH-415-U IR LED (Black LED) pay attention to polarity, press downwards to the PCB
- C1: 100µF at least 16 volt, pay attention to polarity!
- C2, C4: 100nF ceramic capacitor, imprint: 104
- C3: 680pF ceramic capacitor, imprint: 681
- Q1: BC547 (A,B or C) or BC548 (A,B or C)
- R1, R5: 20kΩ_ (red, black, orange, gold)
- R2: 4.7kΩ (yellow, violet, red, gold)
- R3: 470Ω (yellow, violet, black, gold)
- R6: 10kΩ (brown, black, orange, gold)
- R7: 220Ω (red, red, brown, gold)
- TR1: 10KΩ variable resistor
- IC2: SFH5110-36 Infrared receiver IC bends the legs with appropriate tongs! Pay attention to polarity (the curvature must be positioned to the outside)! Caution: electrostatic discharge
- (ESD) and excessive soldering or heating may damage the part!
- X1: 9pol. SUB-D connector, case must be settled close to PCB. Attachment strips must be soldered as well!
- IC1: insert the NE555P, pay attention to polarity!

Results:

When finish building the robot and complete the soldering process, it is satisfying that the robot can work successfully. The results meet the requirements of this experiment completely.

Turning the switch, the robot starts to work though need to take some minutes to run the programs. First of all, the diode12 which is colorless start to shine. When the diode works, the color of the diode will have a change. At the very beginning, the colorless diode becomes yellow for some seconds. After that it will turn to green for other seconds. Finally, the diode will turn off and return to colorless. By finishing the work of the circuit, another diode will work. Diode15 which is red will start to work after diode 12 finish working in some seconds. It will get bright. Next, the three diodes in front of the robot will continue to work when diode12 turns off. The three diodes turn on at the same time at the beginning. Then the two diodes on both sides will get dark. In other words, the middle one gets brighter than the other two diodes. When the process lasts for a short time, the three diodes will turn off. Then the diodes16 will start work, then gets bright red for some seconds.

After all the diodes finish working, the robot will begin to run as the route which is decided by the program. The whole process contains four steps. The robot begins to turn left for the first step. Because the program determines that there is just the right wheel working and the left wheel stay still. Next, the robot will run the opposite program which permits the left wheel to spin and the right wheel stays still, then to find that the robot will get back to the position it has been. Then the robot runs the program which allows both wheels work at the same time. As a result, the robot runs forward in a straight line for a distance. Oppositely, the program will make the robot run back at the route which takes to run forward.

Discussion:

Troubleshooting:

1.Soldering

Soldering is the step which behind the preparing the legs. Soldering is a hard thing. Users put a small quantity of solder to the legs and at the same time, melting, the solder will flow into the metal hole until the hole is filled completely.

More over, soldering is also the most important part in the experiment to do the robot. There are two attentions for soldering. Firstly, for the legs, experimenters should n't leave too much copper legs over the circuit board, it is not beautiful, some future, if two copper legs touched, that may cause the exploding. It's dangerous.

Secondly, for the soldering. The perfect weltering hole is the solder filled the hole exactly. If user putting little solder, it may make the circle doesn't work, otherwise, too much looks bad.

2.The Positive and negative parts of component

3.
Diode-- A diode is a two-terminal electronic component that conducts electric current in only one direction. The function of a diode is to allow an electric current to pass in one direction while blocking current in the opposite direction. Thus, if users put the wrong pole to the holes then the function will be opposite.

LED— LED is also one kind of diodes.
Processor—A processor is the brand of the robot. If experimenter assembled it in the wrong way, the robot may didn't do the right work and even didn't work.

4.Some details

Wheel sensors—the tags on the wheels also are significant part. The tags are the odometer. A complete set of adhesive odometer markers is supplied with this kit. Place them on to the 50/10 gear. Using the sensor-markers with 6 black and white segments is recommended for compatibility for compatibility with other ASURO's. Cars do the different work by the difference between light and dark.

5. C-programme
        Every word in one C-programmed is important, a little mistake also can lead to the destroy.

6. check
        If the car doesn't work, check all above.

Conclusion

To sum up, in this experiment allow every student know exactly how to built up a single robot car from an electronic components. On the other hand, students also gain the experience from the soldering, from build and troubleshooting.
Moreover, there were several limitations during this experiments such as, the student soldering in long duration, may cause some components were burned. So in order to decrease the percentage of burned components, students should solder quickly.

Appendix

A. Parts list
Besides a table-tennis ball the following parts are needed to build an
ASURO:

1x Printed circuit board ASURO
2x Motors Type Igarashi 2025-02
1x Diode 1N4001
8x Diodes 1N4148
4x Transistors BC 327/40 or BC 328/40
4x Transistors BC 337/40 or BC 338/40
1x Integrated circuit CD 4081BE
1x Processor ATmega 8L-8PC (preprogrammed)
1x IR-receiver SFH 5110-36
1x IR-LED SFH415-U
2x phototransistors SFH300
3x LEDs 5mm red bright diffuse or asymmetric wide-angle
1x Dual-LED 3mm
2x Side-phototransistors LPT80A
2x Side-LEDs IRL80A
1x Crystal 8MHz
2x Elco 220µF at least 10V RM 3,5/10
4x ceramic capacitors 100nF RM 5,08
2x ceramic capacitors 4,7nF RM 2,54
1x 100Ω 1/4 W 5%
2x 220Ω 1/4 W 5%
4x 470Ω 1/4 W 5%
10x 1KΩ 1/4 W 5%
1x 1KΩ 1/4 W 1%
3x 2KΩ 1/4 W 1%
2x 4,7KΩ 1/4 W 5%
1x 8,2K Ω 1/4 W 1%
1x 10KΩ 1/4 W 1%
1x 12kKΩ 1/4 W 1%
1x 16K Ω 1/4 W 1%
1x 20KΩ 1/4 W 5%
1x 33kKΩ 1/4 W 1%
1x 68K Ω 1/4 W 1%
1x 1M Ω 1/4 W 5%
3x Socket 14 pol.
6x Detector switch
1x Switch (main on/of)
1x Battery holder
1x Battery clip
1x Jumper
1x Jumper pins 2pole RM 2.5
2x Cogwheel 10/50 cogs; 3,1mm drilling hole Module 0,5

2x Cogwheel 12/50 cogs; 3,1mm drilling hole Module 0,5
2x Pinion gear 10 (oder 12) cogs; drilling hole 1,9mm Module 0,5
2x Collar for 3mm-axle
4x Cable binder
1x Cable binder releasable
2x Rubber tires 38mm
2x Messing shaft 42mm long 3mm diameter,
2x Messing shaft 24,5mm lang, 3mm diameter
ca. 15 cm wire red 0,14mm_
ca. 15 cm wire black 0,14mm_
2x Encoder sticker

The additional RS-232-IR-transceiver will be equipped with the following parts:
1x Printed circuit board IR-RS232-transceiver
3x Diodes 1N4148
1x Zener diode ZPD5.1
1x Transistor BC547 A,B or C or BC548 A,B or C
1x Integrated circuit NE555N
1x IR-receiver SFH 5110-36
1x IR-LED SFH415-U
1x Elco 100µF at least 16V RM 2,5/6
2x ceramic capacitors 100nF RM 5,08
1x ceramic capacitors 680pF RM 2,54
1x 220Ω 1/4 W 5% or better
1x 470Ω 1/4 W 5% or better
1x 4,7KΩ 1/4 W 5% or better
1x 10KΩ 1/4 W 5%
2x 20KΩ 1/4 W 5% or better
1x Trimmer 10k vertically mounted RM 2,5/5
1x Socket 8 pole
1x 9-pol. SUB-D-entry

B. Asuro Diagram

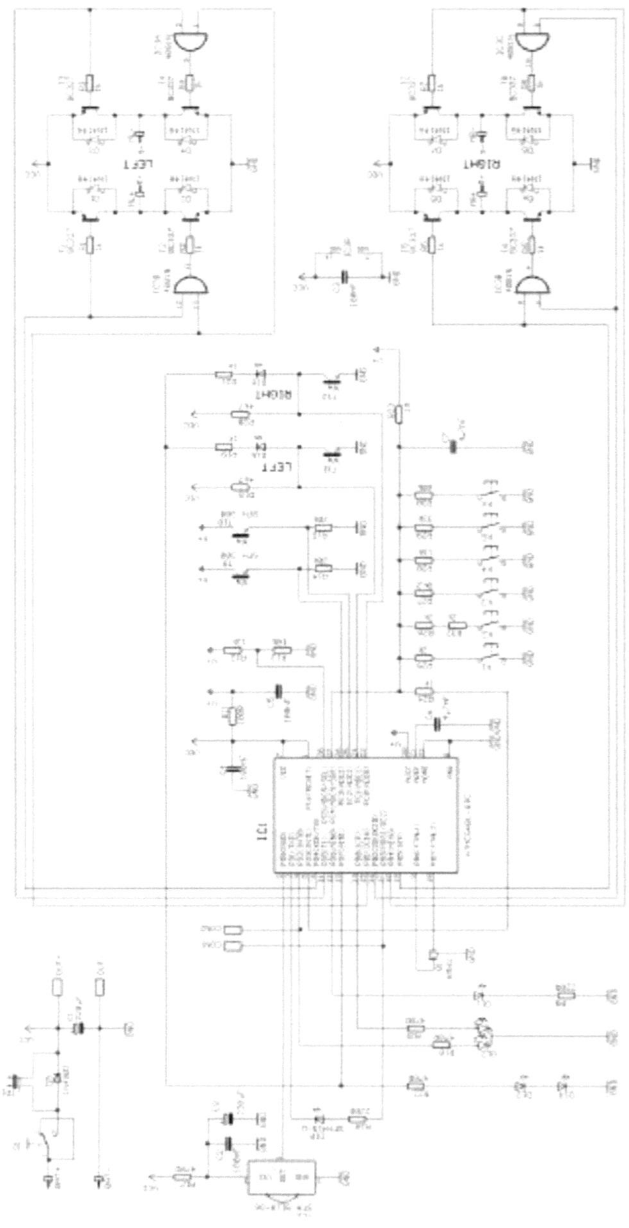

C. RS-232 Transceiver

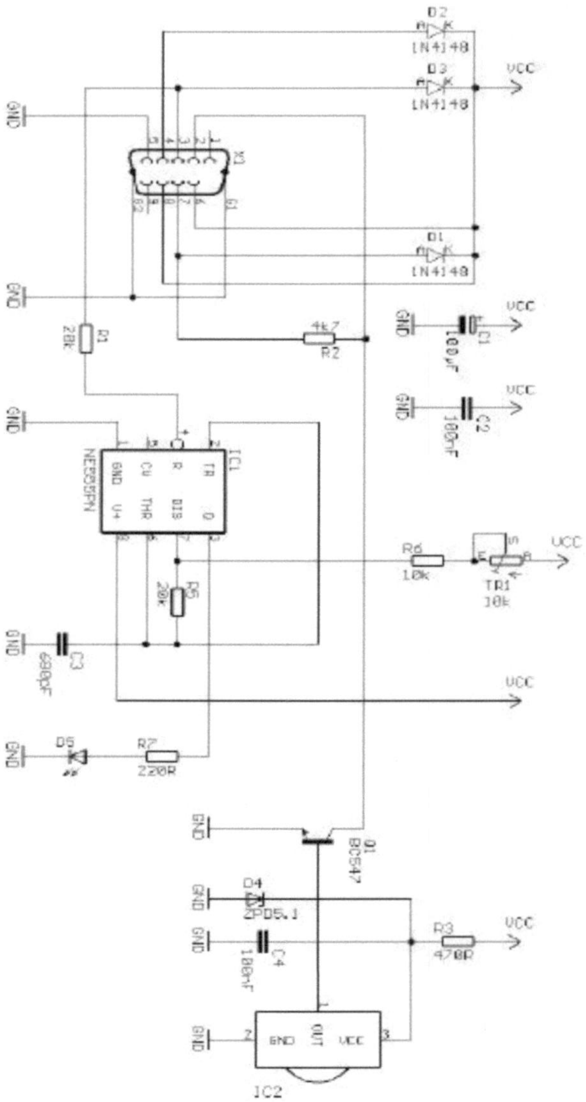

D. Block Diagram Asuro

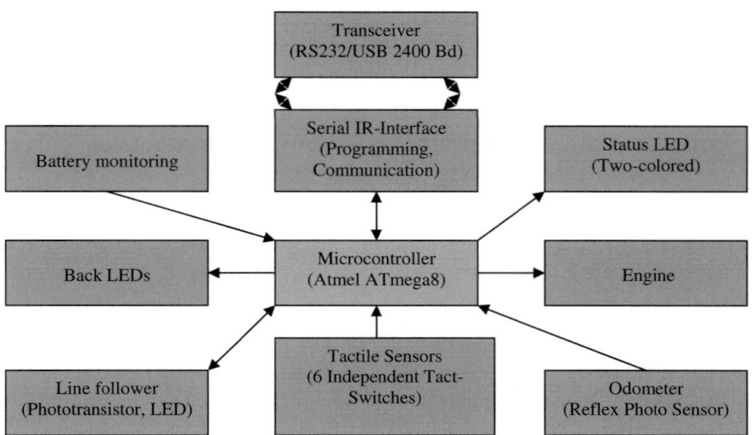

E. Block Diagram PIC Processor

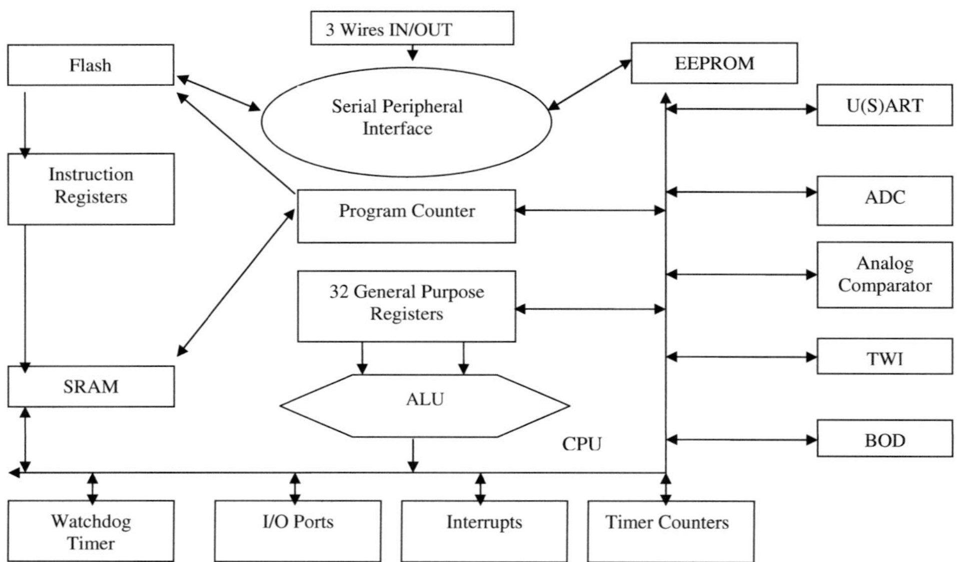

F. Contents Asuro Kit

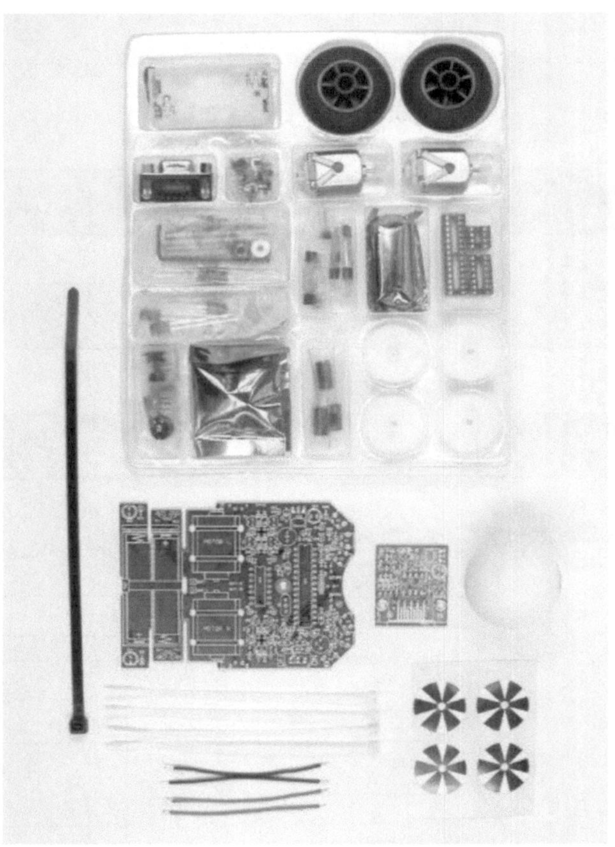

References

"Robot", Wikipedia foundation Inc. accessed November 22, 2010,
<http://en.wikipedia.org/wiki/Robot>

"Educational Robots", Wikipedia foundation Inc., accessed November 22, 2010,
<http://en.wikipedia.org/wiki/Category:Educational_robots>

"Frontiers of Mechanical Engineering in China", Review Article, Springer accessed November 22, 2010
<http://www.springerlink.com/content/c7051m07pt38p242/>

C. Zhao ., 2010, Lectures in experimental skills, Suzhou, China.

YOUR KNOWLEDGE HAS VALUE

- We will publish your bachelor's and
  master's thesis, essays and papers

- Your own eBook and book -
  sold worldwide in all relevant shops

- Earn money with each sale

Upload your text at www.GRIN.com
and publish for free